POSTMODERN ENCOUNTERS

Hawking and the
Mind of God

Peter Coles

Series editor: Richard Appignanesi

ICON BOOKS UK

TOTEM BOOKS USA

Published in the UK in 2000
by Icon Books Ltd., Grange Road,
Duxford, Cambridge CB2 4QF
email: info@iconbooks.co.uk
www.iconbooks.co.uk

Published in the USA in 2000
by Totem Books
Inquiries to: PO Box 223,
Canal Street Station,
New York, NY 10013

Distributed in the UK, Europe,
Canada, South Africa and Asia
by the Penguin Group:
Penguin Books Ltd.,
27 Wrights Lane,
London W8 5TZ

In the United States,
distributed to the trade by
National Book Network Inc.,
4720 Boston Way, Lanham,
Maryland 20706

Published in Australia in 2000
by Allen & Unwin Pty. Ltd.,
PO Box 8500, 9 Atchison Street,
St. Leonards, NSW 2065

Library of Congress catalog
card number applied for

Series editor: Richard Appignanesi

ISBN 1 84046 124 1

Typesetting by Wayzgoose

Printed and bound in the UK by
Cox & Wyman Ltd., Reading

The Hawking Phenomenon

The British theoretical physicist Stephen Hawking is one of the few scientists ever to have become a media celebrity. His book *A Brief History of Time* was a world-wide bestseller, and he has made many appearances on television, not – as with most scientists – restricted to science documentaries, but also in commercials and elsewhere. He is instantly recognisable through the constraints imposed on him by grievous illness. Suffering from the progressive effects of motor neurone disease (amyotrophic lateral sclerosis) since his early twenties, when he was a student at Cambridge, he is confined to a wheelchair and virtually unable to move except to control the computer attached to it. As a result of a tracheotomy in the mid-1980s, he is unable to speak. He talks through an eerily mechanical voice produced by a speech synthesizer.

It is remarkable that a man trapped in a tortured body and deprived of the most basic means of communication could have achieved such fame. And more remarkable still that he has been able to overcome the immense obstacles placed before him to pursue a spectacular career as one of the most imaginative and influential scientists of modern times.

But there is much more to the Hawking phenomenon than his scientific endeavours. The story of 20th-century physics contains many great intellectual achievements by men such as Paul Dirac, Richard Feynman, Erwin Schrödinger and, of course, Albert Einstein. Of this array of geniuses, Albert Einstein is the only household name. In 1998 *The Observer* newspaper named Hawking the 68th most powerful man in Great Britain, as a measure of his impact on people's daily lives. As far as the media and, perhaps, popular consciousness are concerned, Hawking and Einstein rank at a similar level in the physicists' hall of fame. Indeed, when Stephen Hawking appeared in *Star Trek: The Next Generation* (playing himself), it was alongside Sir Isaac Newton and Einstein.

What I want to do in this book is to try to find a proper perspective in which to view the work of Stephen Hawking and his place in wider society. My argument is that there is much more to the

Hawking phenomenon than either his scientific achievements or the sympathy engendered by his illness, or even a combination of these two. The extra ingredient can only be seen by viewing Hawking in the context of the tremendous changes not only in science itself, but also in the relationship between man and nature, that have taken place in the last 100 years or so.

The title comes from the last sentence of *A Brief History of Time* (1988), in which Hawking writes of his desire to 'know the Mind of God'. This phrase is key to understanding the wider role of Hawking beyond the rarefied world of abstract mathematical theory. To see why, we have to explore the development of physics from the beginning of the modern era.

A Brief History of Physics

By a curious numerological coincidence, Stephen Hawking was born 300 years to the day after the death of Galileo Galilei, the man who did most to usher in the era of modern science. But for the purposes of this story, it is best to start with Sir Isaac Newton, who was the first truly mathematical physicist and thus a direct ancestor of Stephen Hawking. The first great achievement of theoretical physics was Newton's theory of mechanics (see 'Key Ideas' at the end of this book), which is encoded in three simple laws that are probably still remembered even by those who haven't studied physics since their school days:

(i) Every body continues in a state of rest or uniform motion in a straight line unless it is compelled to change that state by forces impressed upon it.

(ii) Rate of change of momentum is proportional to the impressed force, and is in the direction in which this force acts.

(iii) To every action, there is always opposed an equal reaction.

These three laws of motion are general, applying just as accurately to the behaviour of balls on a billiard table as to the motion of the planets around the Sun. It was Newton's insight into the problem of planetary motion that it could be described by the same mathematical law as objects on Earth, such as apples falling from a tree. Newton realised that a body orbiting in a circle, like the Moon going around the Earth, is experiencing a force in the direction of the centre of motion (just as a weight tied to the end of a piece of string does when it is twirled around one's head). An apple feels a downward force towards the centre of the Earth. Based on this idea, Newton developed a theory of Universal Gravitation that could explain the motion of the planets discussed by Johannes Kepler more than a century earlier. This was the first proper example of apparently disparate phenomena being *unified*, i.e., incorporated in a single mathematical theory.

The idea of a universe governed by Newton's laws of motion was to dominate scientific thinking for more than two centuries. But wider than that, Newton's achievements suggested a perfectly pre-dictable cosmos whose behaviour was as regular as clockwork. Once one knew the state of the Solar System at any time, one could predict its state at any time in the future with total confidence. Newton, a profoundly but unconventionally religious man, had unwittingly changed the role of God. Instead of intervening in the daily running of the world, He simply had to wind it up and let it go.

This view of a rigidly predictable universe was to hold sway until the end of the 19th century. But in the meantime, other branches of science came under the scrutiny of mathematical physicists inspired by Newton's example. Chief among these was the theory of electricity and magnetism. It was known that objects could be charged and that objects of opposite charge tend to attract each other, while particles of the same charge tend to repel. Coulomb's law of electrostatics, which accounted for these phenomena, was very similar to Newton's law of gravitation. Michael Faraday (through no fault of his own, Margaret Thatcher's favourite scientist) had done marvellous experimental work which showed that electricity and magnetism were related in some way. Moving charges generate magnetism, which in the early history of physics was thought to be a different kind of phenomenon altogether. James Clerk Maxwell was the first to elucidate the character of these interactions – now known as electromagnetic interactions – in a set of mathematical laws known as Maxwell's equations. These showed further that electricity and magnetism could fluctuate together in waves that travel at the speed of light. This led to the realisation that light was a form of electromagnetic wave, and that other forms of electromagnetic wave would be possible (such as radio waves).

So successful was this programme that physicists at the end of the 19th century were filled with confidence that soon all physical phenomena would surrender to a Newtonian treatment. This confidence was soon to be shattered.

The World of the Quantum

The early years of the 20th century saw two revolutions in physics. The first of these was the birth of quantum mechanics. It changed forever the mechanistic view of a world founded upon Newton's laws of motion. As I have already mentioned, a universe running according to Newtonian physics is *deterministic*, in the sense that if one knew the positions and velocities of all the particles in a system at a given time, then one could predict their behaviour at all subsequent times. Quantum mechanics changed all that, since one of the essential components of this theory is the principle (now known as Heisenberg's Uncertainty Principle) that, at a fundamental level, the behaviour of particles is inherently unpredictable.

In the world according to quantum theory, every entity has a dual nature. In classical physics, two distinct concepts were used to describe distinct natural phenomena: waves and particles. Quantum physics tells us that these concepts do not apply separately to the microscopic world. Things that we previously imagined to be particles (point-like objects) can sometimes behave like waves. Phenomena that we previously thought of as waves can sometimes behave like particles. For example, light can behave like a wave phenomenon – one can display interference and diffraction effects using prisms and lenses. Moreover, Maxwell had shown that light was actually described mathematically by an equation called the wave equation. The wave nature of light is therefore predicted by this theory.

On the other hand, Max Planck's work on the

radiation emitted by hot bodies had also shown that light could behave as if it came in discrete packets, which he called *quanta*. He hesitated to claim that these quanta could be identified with particles. It was in fact Albert Einstein, in his work on the photoelectric effect for which he won the Nobel Prize, who made the step of saying that light was actually made of particles. These particles later became known as photons.

So how can something be both a wave and a particle? One has to say that reality cannot be exactly described by either concept, but that it behaves sometimes as if it were a wave and sometimes as if it were a particle.

Imagine a medieval monk returning to his monastery after his first trip to Africa. During his travels he chanced upon a rhinoceros, and is now faced with the task of describing it to his incredulous brothers. Since none of them has ever seen anything as strange as a rhino in the flesh, he has to proceed by analogy. The rhinoceros, he says, is in some respects like a dragon and in others like a unicorn. The brothers then have a reasonable picture of what the beast looks like. But neither dragons nor unicorns exist in nature, while the rhinoceros does. It is the same with our quantum world. Reality is described neither by idealised waves nor by idealised particles, but these concepts can give some impression of certain aspects of the way things really are.

The idea that energy came in discrete packets (or quanta) was also successfully applied to the simplest of all atoms – the hydrogen atom – by Niels Bohr in 1913, and to other aspects of atomic and nuclear physics. The existence of discrete energy levels in atoms and molecules is fundamental to the field of spectroscopy, which plays a role in areas as diverse as astrophysics and forensic science.

The Uncertain Universe

But the acceptance of the quantised nature of energy (and light) was only the start of the revolution that founded modern quantum mechanics. It was not until the 1920s and the work of Erwin Schrödinger and Werner Heisenberg that the dual nature of light as both particle and wave was finally elucidated. For while the existence of photons had become accepted in the previous years, there had been no way to reconcile this with the well-known wave behaviour of light. What emerged in the 1920s was a theory of quantum physics built upon wave mechanics. In Schrödinger's version of quantum theory, the behaviour of all systems is described in terms of a wave function which evolves according to an equation called the Schrödinger equation. The wave function depends on both space and time. Just as Maxwell had found for electromagnetism, Schrödinger's equation describes *waves*.

So how does the *particle* behaviour come in? The answer is that the quantum wave function does not describe something like an electro-magnetic field which one thinks of as a physical thing existing at a point in space and fluctuating in time. The quantum wave function describes a *probability* wave. Quantum theory asserts that the wave function is all one can know about the system. One cannot predict with certainty exactly where the particle will be at a given time – just the probability.

An important aspect of this wave-particle

duality is the Uncertainty Principle. This has many repercussions for physics, but the simplest one involves the position of a particle and its speed. Heisenberg's Uncertainty Principle states that one cannot know the position and speed of a particle independently of one another. The better you know the position, the worse you know the speed, and vice-versa. If you can pinpoint the particle exactly, then its speed is completely unknown. If you know its speed precisely, then the particle could be located anywhere. This principle is quantitative and does not apply only to position and momentum, but also to energy and time and other pairs of quantities that are known as 'conjugate variables'. It is a particularly important consequence of the energy-time Uncertainty Principle that empty space can give birth to short-lived particles that spring in and out of existence on a timescale controlled by the Uncertainty Principle.

The interpretation to be put on this probabilistic approach is open to considerable debate. For example, consider a system in which particles travel in a beam towards two closely-separated slits. The wave-function corresponding to this situation displays an interference pattern because the 'probability wave' passes through both slits. If the beam is powerful, it will consist of huge numbers of photons. Statistically the photons should land on a screen behind the slits according to the probability dictated by the wave-function. Since the slits set up an interference pattern, the screen will show a complicated series of bright and faint bands where the waves sometimes add up 'in phase' and sometimes cancel each other. This seems reasonable, but suppose we turn down the power of the beam. This can be done in such a way that there is only one photon at any time travelling through the slits. The arrival of each photon can be detected on the screen. By running the experiment for a reasonably long time, one can build up a pattern on the screen. Despite the fact that only one photon at a time is travelling through the apparatus, the screen still shows the pattern of fringes. In some sense, each photon must turn into a wave when it leaves the source, travel through both slits interfering with itself on the way, and then turn back into a photon in order to land in a definite position on the screen.

So what is going on? Clearly each photon lands in a particular place on the screen. At this point we know its position for sure. What does the wave-function for this particle do at this point? According to one interpretation – the so-called Copenhagen interpretation – the wave-function collapses so that it is concentrated at a single point. This happens whenever an experiment is performed and a definite result is obtained. But before the outcome is settled, nature itself is indeterminate. The photon really doesn't go through either one of the slits – it is in a 'mixed' state. The act of measurement changes the wave-function and therefore changes reality. This has led many to speculate about the interaction between consciousness and quantum 'reality'. Is it *consciousness* that causes the wave-function to collapse?

A famous illustration of this conundrum is provided by the paradox of Schrödinger's Cat. Imagine there is a cat inside a sealed room containing a vial of poison. The vial is attached to a device which will break it and poison the cat when a quantum event occurs, for example the emission of an alpha-particle by a lump of radioactive material. If the vial breaks, death is instantaneous. Most of us would accept that the cat is either alive or dead at a given time. But if one takes the Copenhagen interpretation seriously, it is somehow both. The wave-function for the cat comprises a superposition of the two possible states. Only when the room is opened, and the state of the cat 'measured', does it 'become' either alive or dead.

An alternative to the Copenhagen interpretation is that nothing physically changes at all when a measurement is performed. What happens is that the observer's state of knowledge changes. If one asserts that the wave-function ψ represents what is known by the observer rather than what is true in reality, then there is no problem in having it change when a particle is known to be in a definite state. This view suggests an interpretation of quantum mechanics in which at some level things might be deterministic, but we simply do not know enough to predict.

Yet another view is the Many Worlds interpretation. In this, every time an experiment is performed (e.g. every time a photon passes through the slit device) the universe, as it were, splits into two. In one universe the photon goes through the left-hand slit and in the other it goes through the right-hand slit. If this happens for every photon, one ends up with an enormous number of parallel universes. All possible outcomes of all possible experiments occur in this ensemble. But before I head off into a parallel universe, let me resume the thread of the story.

The Relativity Revolution

The second shattering blow to the edifice of 19th-century physics was the introduction, by Albert Einstein, of the principle of relativity. The idea of relativity did not originate with Einstein, but dates back at least as far as Galileo. Galileo claimed that *relative motion* matters, so there could be no such thing as *absolute motion*. He argued that if one were travelling in a boat at constant speed on a smooth lake, then there would be no experiment that one could do in a sealed cabin on the boat that would indicate that one was moving at all. Of course, not much was known about physics in Galileo's time, so the kinds of experiment he could envisage were rather limited.

Einstein's version of the principle of relativity simply turned it into the statement that all laws of nature have to be exactly the same for all observers in relative motion. In particular, Einstein decided that this principle must apply to the theory of electromagnetism, constructed by James Clerk Maxwell, which describes amongst other things the forces between charged bodies mentioned above. One of the consequences of Maxwell's theory is that the speed of light (in vacuum) appears as a universal constant (usually called 'c'). Taking the principle of relativity seriously means that all observers have to measure the same value of c, whatever their state of motion. This seems straightforward enough, but the consequences are nothing short of revolutionary.

Einstein decided to ask himself specific questions about what would be observed in particular kinds of experiments involving the exchange of light signals. Einstein worked a great deal with *gedanken* (thought) experiments of this kind. For example, imagine there is a flash bulb in the centre of a railway carriage moving along a track. At each end of the carriage there is a clock, so that when the flash illuminates it we can see the time. If the flash goes off, then the light signal reaches both ends of the carriage simultaneously from the point of view of someone sitting in the carriage: the same time is seen on each clock.

Now picture what happens from the point of view of an observer at rest who is watching the train from the track. The light flash travels with the same speed in our reference frame as it did for the passengers. But the passengers at the back of the carriage are moving into the signal, while those at the front are moving away from it. The observer on the track therefore sees the clock at the back of the train light up before the clock at the front does. But when the clock at the front does light up, it reads the same time as the clock at the back did! This observer has to conclude that something is wrong with the clocks on the train.

This example demonstrates that the concept of simultaneity is relative. The arrivals of the two light flashes are simultaneous in the frame of the carriage, but occur at different times in the frame of the track.

What Einstein had shown was that the foundations of Newton's laws of motion were shaky. Newton had assumed that one could give an absolute meaning to the distance between two objects, as if they were located on a piece of God-given graph paper. He had also assumed that there was an absolute time that ticked at the same rate for all observers regardless of their state of motion. Ideas of space and time are embedded deeply in Newton's three laws of motion. How can one talk about whether a body is at rest, if one does not say who is watching it and how they are moving?

In relativity theory, it is not helpful to think of space and time as separate things, because they are not absolute in themselves. It is possible, however, to construct a kind of generalisation of three-dimensional space that incorporates time as one facet. The idea of a four-dimensional *space-time* accomplishes this.

Einstein's special theory of relativity was a remarkable achievement, but he did not stop there.

He spent the next ten years working on a complete generalisation of the theory that would enable him fully to overthrow the Newtonian conception of the world by replacing Newton's law of Universal Gravitation. Einstein's general theory of relativity is essentially his theory of gravity. The relativity of time embodied in the special theory is present in the general theory, but there are additional effects of time dilation and length contraction due to gravitation. Space-time becomes warped by the presence of gravity, light ceases to travel in straight lines and even time-travel becomes a (perhaps purely mathematical) possibility.

The rise of general relativity is an interesting tale in itself, and is described in my previous book in this series, *Einstein and the Total Eclipse* (to be re-issued in autumn 2000 as *Einstein and the Birth of Big Science*).

The Four Forces of Nature

Armed with the new theories of relativity and quantum mechanics, and in many cases further spurred on by new discoveries made possible by advances in experimental technology, physicists in the 20th century sought to expand the scope of science to describe the natural world. All phenomena amenable to this treatment can be attributable to the actions of the four forces of nature. These four fundamental interactions are the ways in which the various elementary particles from which all matter is made interact with each other. Two of these four I have already discussed: electromagnetism and gravity. The other two concern the interactions between the constituents of the nuclei of atoms, the weak nuclear force and the strong nuclear force. The four forces vary in strength (gravity is the weakest and the strong nuclear force is the strongest), and also in the kinds of elementary particles that take part in the interactions they control.

The electromagnetic force holds electrons in orbit around atomic nuclei, and is thus responsible for holding together all material with which we are familiar. However, it was realised early in the 20th century that, in order to apply Maxwell's theory in detail to atoms, ideas from quantum physics and relativity would have to be incorporated. It was not until the work of Richard Feynman and others, building on the work of Dirac, that a full quantum theory of the electromagnetic force, called quantum electrodynamics, was developed. In this theory, usually abbreviated to QED, electromagnetic radiation in the form of photons is responsible for carrying the electromagnetic interaction between particles of different charges.

The next force to come under the spotlight was the weak nuclear force, which is responsible for the decay of certain radioactive material. It involves a particular class of elementary particles called the leptons, of which the best-known example is the electron. As in the case of electromagnetism, weak forces between particles are mediated by other particles – not photons, in this case, but massive particles called the W and Z bosons. Photons have no mass, and so have a long range, but W and Z bosons *have* mass and this reduces their range, so the effects of the weak force are confined to the tiny scales of an atomic nucleus. The W and Z particles otherwise play the same role in this context as the photon does in QED. They, and the photon, are examples of what are known as gauge bosons.

The strong nuclear interaction (or strong force) involves another family of elementary particles called the hadrons, which includes the protons and neutrons that make up the nuclei of atoms. The theory of these interactions is called quantum chromodynamics (or QCD), and it is built upon similar lines to the theory of the weak inter-action. In QCD, there is another set of gauge bosons to mediate the force. These are called gluons, and they interact in a different way from the W and Z particles. This means that the strong force is of even shorter range than the weak force. There are eight gluons, and playing a similar role to that of electric charge in QED, there is a prop-erty called 'colour'. The hadrons are represented as collections of 'quarks' which have a fractional electrical charge and come in six different 'flavours': up, down, strange, charmed, top, and bottom. Each distinct hadron species is a differ-ent combination of the quark flavours.

These theories all successfully unite quantum physics with Einstein's special theory of relativity. The one force of nature which is left out, is gravity. There is no theory that puts gravity and quantum mechanics together in a mathematically consistent way. Not yet, anyway.

The Drive for Unification

The achievements of theoretical physics did not stop with the elucidation of the quantum theories of electromagnetism and the weak and strong nuclear interactions. Would it be possible, taking a cue from Maxwell's original unification, to put all three of these forces together in a single overarching theory?

A theory that unifies the electromagnetic force with the weak nuclear force was developed around 1970 by Glashow, Salam and Weinberg. Called the electroweak theory, this represents these two distinct forces as being the low-energy manifestations of a single force. When particles have low energy, and are moving slowly, they feel the different nature of the weak and electromagnetic forces. Physicists think that at high energies there is a symmetry between the electromagnetic and weak interactions. Electromagnetism and the weak force appear different to us at low energies because this symmetry is broken. Imagine a pencil standing on its end. When vertical, it looks the same from all directions. A random air movement or a passing lorry will cause it to topple – it could fall in any direction with equal probability. But when it falls, it falls some *particular* way defining a particular direction. In the same way, the difference between electromagnetism and weak nuclear forces could be just happenstance, a chance consequence of how the high-energy symmetry was broken.

The electroweak and strong interactions co-exist in a combined theory of the fundamental interactions called 'the standard model'. This model is, however, not really a unified theory of all three interactions in the same way that the electroweak theory is for two of them. Physicists hope eventually to unify all three of the forces discussed so far in a single theory, which would be known as a grand unified theory, or GUT. There are many contenders for such a theory, but it is not known which (if any) is correct.

The fourth fundamental interaction is gravity, and the best theory of it is general relativity. This force has proved extremely resistant to efforts to make it fit into a unified scheme of things. The first step in doing so would involve incorporating quantum physics into the theory of gravity in order to produce a theory of quantum gravity. Despite strenuous efforts, this has not yet been achieved. If it is ever done, the next task will be to unify quantum gravity with the grand unified theory. The result of this endeavour would be, according to Stephen Hawking, 'A Theory of Everything', and knowing it would be like knowing the Mind of God.

The Missing Link

It always seems to me quite ironic that it is gravity, which really began the modern era of theoretical physics, that should provide the stumbling-block to further progress towards a unified theory of all the forces of nature. In many ways, the force of gravity is extremely weak. Most material bodies are held together by electrical forces between atoms which are many orders of magnitude stronger than the gravitational forces between them. But despite its weakness, gravity has a perplexing nature that seems to resist attempts to put it together with quantum theory.

The best theory of gravity that seems to be available at the present time is Einstein's general theory of relativity. This is a classical theory, in the sense that Maxwell's equations of electromagnetism are also classical, in that they both involve entities that are smooth rather than discrete, and describe behaviour that is deterministic rather than probabilistic. On the other hand, quantum physics describes a fundamental lumpiness – everything consists of discrete packets or quanta. Likewise, the equations of general relativity allow one to calculate the exact state of the universe at a given time in the future, if sufficient information is given at some time in the past. They are therefore deterministic. The quantum world, on the other hand, is subject to the uncertainty embodied in Heisenberg's Uncertainty Principle.

Of course, classical electromagnetic theory is perfectly adequate for many purposes, but it does break down in certain situations, such as when radiation fields are very strong. For this reason, physicists sought (and eventually found) the quantum theory of electromagnetism or quantum electrodynamics (QED). This theory was also made consistent with the special theory of relativity, but does not include general-relativistic effects.

While Einstein's equations also seem quite accurate for most purposes, it is similarly natural to attempt the construction of a quantum theory of gravity. Einstein himself always believed that his theory was incomplete in this sense, and would eventually need to be replaced by a more complete theory. By analogy with the breakdown of classical electromagnetism, one can argue that this should happen when gravitational fields are very strong, or on length scales that are extremely short. Attempts to build such a theory have been largely unsuccessful, mainly for complicated technical reasons to do with the mathematical construction of Einstein's theory.

Although there is nothing resembling a complete picture of what a quantum theory of gravity might involve, there are some interesting speculative ideas. For example, since general relativity is essentially a theory of space-time, space and time themselves must become quantised in quantum gravity theories. This suggests that, although space and time appear continuous and smooth to us, on minuscule scales equivalent to the Planck length (around 10^{-33}cm), space is much more lumpy and complicated, perhaps consisting of a foam-like topology of bubbles connected by tunnels called wormholes that are continually forming and closing again on a timescale corresponding to the Planck time, which is 10^{-43} seconds. It also seems to make sense to imagine that quantised gravitational waves, or gravitons, might play the role of the gauge bosons in other fundamental interactions, as do the photons in the theory of quantum electrodynamics. As yet, there is no concrete evidence that these ideas are correct.

The tiny scales of length and time involved in quantum gravity demonstrate why this is a field for theorists rather than experimentalists. No device has yet been built capable of forcing particles into a region equivalent to the Planck length or less. Attempts to figure out what happens when quantum physics and gravity go together have to involve regions where gravity is so strong that its quantum nature reveals itself. It is at this inter-face that Stephen Hawking's work is situated. Having described the context, I will now give a brief overview of some of his major contributions.

Black Holes . . .

A black hole is a region of space-time in which the action of gravity is sufficiently strong that light cannot escape. The idea that such a phenomenon might exist dates back to John Michell, an English clergyman, in 1783, but black holes are most commonly associated with Einstein's theory of general relativity. Indeed, one of the first exact solutions of Einstein's equations to be obtained mathematically describes such an object.

Karl Schwarzschild obtained his famous solu-tion of the Einstein equations in 1916, only a year after the publication of Einstein's theory. He died soon afterwards on the Eastern front during the First World War. The solution corresponds to a spherically-symmetric distribution of matter, and it was originally intended that this could form the basis of a mathematical model for a star. It was soon realised, however, that for an object of any mass, the Schwarzschild solution implied the existence of a critical radius (now called the Schwarzschild radius). If a massive object lies entirely within its Schwarzschild radius, then no light can escape from the surface of the object. For the mass of the Earth, the critical radius is only 1cm, whereas for the Sun it is about 3km. So to make a black hole out of the Sun would require compressing the solar material to a phenomenal density.

Since the pioneering work of Schwarzschild, research on black holes has been intense. Although there is as yet no watertight evidence for the existence of black holes in nature, they are thought to exist in many kinds of astronomical object. The intense gravitational field surrounding a black hole of about 100 million times the mass of the Sun is thought to be the engine that drives the enormous luminosity of certain types of galaxies. More recent observational studies of the dynamics of stars near the centres of galaxies indicate very strong mass concentrations that are usually identified with black holes with masses similar to this figure. Black holes of much smaller mass may be formed at the end of a star's life, when its energy source fails and it collapses in on itself. It is also possible that very small black holes, with masses ranging from millions of tonnes to less than a gramme, might have been formed very early on in the Big Bang. Such objects are usually called primordial black holes.

As well as having potentially observable consequences, black holes also pose fundamental questions about the applicability of general relativity. In this theory, the inability of light to escape is due to the extreme curvature of space-time. It is as if the space around the hole is wrapped up into a ball so that light travels around the surface of the ball, but cannot escape.

. . . and why they ain't so black

Technically, the term 'black hole' actually refers to a thing called the 'event horizon' that forms around the collapsed object, causing the space-time distortion. The horizon defines the edge of the hole; no communication is possible between the regions of space-time inside the horizon and outside. The presence of the event horizon ensures that no light or other form of radiation can escape from the black hole, which therefore seems an entirely appropriate name.

But this common-sense picture of black holes was refuted in the 1970s by calculations performed by Stephen Hawking. Hawking was interested in trying to explore the consequences of quantum physics in regions where the gravitational fields are strong. A black hole is a good place to try. Hawking showed that, under certain circumstances, black holes could emit radiation; in fact, that they could emit so much radiation that in the end they should evaporate entirely. The radiation emitted by black holes in this way is now called Hawking radiation.

How can radiation be emitted by a black hole when the hole is surrounded by a horizon? The reason why Hawking radiation is allowed is that it is essentially a quantum process. Nothing described by classical physics can leave the horizon of a black hole, but this does not necessarily apply to all circumstances when quantum physics is involved. The violation of classical restrictions occurs in many other situations where quantum phenomena involve 'tunnelling', such as the ability of elementary particles to escape from situations where one might expect them to be trapped by electromagnetic forces.

It works in the following way. The space-time around a black hole may be represented as a vacuum, but this vacuum is not entirely empty. Tiny quantum fluctuations are continually forming and decaying according to Heisenberg's Uncertainty Principle. What this means is that the vacuum is filled with ephemeral particles that form and decay on a very short timescale – these are called virtual particles. Such particles always form as a particle and an anti-particle. Matter and anti-matter annihilate when brought together, so the pair of particles exists only for an instant while the chance nature of quantum mechanics allows them briefly to separate. Usually a particle/anti-particle pair is created *ex nihilo* (from nothing) by such processes, but the pair never separates very far and the two particles annihilate each other shortly after they form. On the edge of a horizon, however, even a small separation can be crucial. If one particle of the pair happens to move inside the horizon, it is lost forever, while the other particle escapes. To all purposes, this looks like the black hole is radiating from its event horizon. The hole is therefore not completely black. Hawking also showed that the mass of the black hole goes down while this happens.

The smaller the black hole, the more efficient is this process. It only really has observable consequences for black holes that are very small indeed. The temperature at the surface of a black hole depends inversely on the mass of the hole. Smaller-mass black holes therefore evaporate more quickly than large ones, and produce a much higher temperature. But evaporation is the fate of all black holes. They glow dimly at first, but as they fritter away their mass they glow more brightly. The less massive they get, the hotter they get, and the more quickly they lose mass. Eventually, when they get very small indeed, they explode in a shower of high-energy particles.

This effect is particularly important for very small black holes which, in some theories, form in the very early universe – the so-called primordial black holes. Any such objects with a mass less than about 10^{15} grammes (about the mass of a mountain) would have evaporated by now, and the high-energy gamma rays they produced may well be detectable. The fact that this radiation is not observed places strong constraints on theories that involve these primordial objects.

Nobody has yet observed Hawking radiation, but although Hawking's calculation was initially disputed, it is now recognised to be a correct application of quantum theory to one aspect of the behaviour of matter in a strong gravitational field. As such, it is one step towards a theory of quantum gravity.

The Singular Nature of Gravity

Hawking's calculations had shown that interesting things can happen around the edges of a black hole, near the horizon. But what happens inside the horizon? According to the famous theorems of Roger Penrose and others, the inevitable result is not nice. It is a 'singularity'.

In mathematics, a singularity is a pathological property wherein the value of a particular quantity becomes infinite during the course of a calculation. To give a very simplified example, consider the calculation of the Newtonian force due to gravity exerted by a massive body on another particle. This force is inversely proportional to the square of the distance between the two bodies, so that if one tried to calculate the force for objects at zero separation, the result would be infinite. Singularities are not always signs of serious mathematical problems. Sometimes they are simply caused by an inappropriate choice of coordinates. For example, something strange and akin to a singularity happens in the standard maps to be found in an atlas. These maps look quite sensible until you look very near the poles. In a standard equatorial projection, the North Pole does not appear as a point as it should, but is spread out from a point to a straight line along the top of the map. But if you were to travel to the North Pole you would not see anything catastrophic there. The singularity that causes this point to appear is an example of a coordinate singularity, and it can be transformed away by using a different kind of projection. Nothing particularly odd will happen to you if you attempt to cross this kind of singularity.

Singularities occur with depressing regularity in solutions of the equations of general relativity. Some of these are coordinate singularities like the one discussed above, and are not particularly serious. However, Einstein's theory is special in that it predicts the existence of real singularities where real physical quantities that should know better, such as the density of matter or the temperature, become infinite. The curvature of space-time can also become infinite in certain situations. The existence of these singularities suggests to many that some fundamental physics, describing the gravitational effect of matter at extreme density, is absent from our understanding. It is possible that a theory of quantum gravity might enable physicists to calculate what happens deep inside a black hole without having all mathematical quantities becoming infinite. Indeed, Einstein himself wrote in 1950:

The theory is based on a separation of the concepts of the gravitational field and matter. While this may be a valid approximation for weak fields, it may presumably be quite inadequate for very high densities of matter. One may not therefore assume the validity of the equations for very high densities and it is just possible that in a unified theory there would be no such singularity.

Probably the most famous example of a singularity lies at the centre of a black hole. This appears in the original Schwarzschild solution corresponding to a hole with perfect spherical symmetry. For many years, physicists thought that the existence of a singularity of this kind was merely due to the rather artificial special nature of this spherical solution. However, a series of mathematical investigations, culminating in the singularity theorems of Penrose that I have mentioned above, showed that no special symmetry is required and that singularities arise whenever any objects collapse under their own gravity.

As if to apologise for predicting these singularities in the first place, general relativity does its best to hide them from us. A Schwarzschild black hole is surrounded by an event horizon that effectively protects outside observers from the singularity itself. It seems likely that all singularities in general relativity are protected in this way, and so-called 'naked singularities' are not thought to be physically realistic.

The Bug in the Big Bang

Roger Penrose's work on mathematical properties of the black hole singularity made a big impression on the young Stephen Hawking, who became interested in the problem of trying to apply them elsewhere. Penrose had considered what would happen in the future when an object collapsed under its own gravity. Hawking was interested to know whether these ideas could be applied instead to the problem of understanding what had happened in the *past* to a system now known to be expanding, i.e., the universe! Hawking contacted Roger Penrose about this, and they worked together on the problem of the Big Bang singularity, as it is now known.

The Big Bang furnishes the standard theoretical framework through which cosmologists interpret observations and construct new theoretical ideas. It is not entirely correct to call the Big Bang a 'theory'. I prefer to use the word 'model'. The difference between theory and model is subtle, but a useful definition is that a theory is usually expected to be completely self-contained (it can have no adjustable parameters, and all mathematical quantities are defined *a priori*), whereas a model is not complete in the same way. According to the Big Bang model, the universe originated from an initial state of high temperature and density, and has been expanding ever since. The dynamics of the Big Bang are described mathematically by the equations of Einstein's theory of general relativity. These models predict the existence of a singularity at the very beginning, where the temperature and density are infinite. Since this event is the feature that best encapsulates the nature of the model, many people use the phrase 'Big Bang' to refer to the very beginning, rather than to the subsequent evolution of the universe.

Most cosmologists interpret the Big Bang singularity in much the same way as the black hole singularity discussed above, i.e., as meaning that Einstein's equations break down at some point in the early universe due to the extreme physical conditions present there. If this is the case, then the only hope for understanding the early stages of the expansion of the universe is through a theory of quantum gravity. Since we don't have such a theory, the Big Bang is incomplete. One can estimate the scales of length and time for which this happens. Our understanding of the universe breaks down completely for times before the Planck time, which is about 10^{-43} seconds after the Big Bang itself.

This shortcoming is the reason why the word 'model' is probably more appropriate than 'theory' for the Big Bang. The problem of not knowing about the initial conditions of the universe is the reason why cosmologists still cannot answer some basic questions, such as whether the universe will expand forever or eventually re-collapse.

Within the basic framework of a cosmological model, laws of physics known from laboratory experiments or assumed on the basis of theoretical ideas can be used to infer the physical conditions at different stages of the expansion of the universe. In this way, the thermal history of the universe is mapped out. The further into the past we extrapolate, the hotter the universe gets and the more exotic is the physical theory required. With present knowledge of the physics of elementary particles and fundamental interactions, cosmologists can turn the clock back from the present age of the universe (some 15 billion years or so) and predict with reasonable confidence what was happening within about a microsecond of the Big Bang. Using more speculative physical theory not tested in the laboratory, including grand unified theories, cosmologists have tried to push the model to within 10^{-35} seconds of the very beginning, leading to refinements and extensions of the basic picture. One particular aspect of this, in which Hawking himself has played a role, is in the idea that the universe may have undergone a period of accelerated expansion very early on, leading to a model called 'cosmic inflation'.

Despite gaps in our knowledge of physics at the very highest energies, the theory is widely accepted, principally because it accounts for the following three observations:

(i) the expansion of the universe, as discovered by Hubble in 1929;

(ii) the existence of the cosmic microwave background radiation (a relic of the primordial hot phase of the universe, discovered by Penzias and Wilson in 1965);

(iii) the abundances of the light chemical elements hydrogen, helium and lithium, produced by nuclear fusion in the primordial fireball.

Cleansing the Infinities

The presence of a singularity at the very beginning of the universe is very bad news for the Big Bang model. Like the black hole singularity, it is a real singularity where the temperature and density become truly infinite. In this respect, the Big Bang can be thought of as a kind of time-reverse of the gravitational collapse that forms a black hole. As was the case with the Schwarzschild solution, many physicists thought that the initial cosmological singularity could be a consequence of the special form of the solutions of Einstein's equations used to model the Big Bang, but this is now known not to be the case. Hawking and Penrose generalised Penrose's original black hole theorems to show that a singularity invariably exists in the past of an expanding universe in which certain very general conditions apply. Physical theory completely fails us at the instant of the Big Bang, where the nasty infinities appear.

So is it possible to avoid this singularity? And if so, how? The most likely possibility is that the initial cosmological singularity might well just be a consequence of extrapolating deductions based on the classical theory of general relativity into a situation where this theory is no longer valid. This is what Einstein says in the paragraph quoted above during the discussion of black holes. What is needed is quantum gravity, but we don't have such a theory. And, since we don't have it, we don't know whether it would solve the riddle of the universe's apparently pathological birth.

There are, however, ways of avoiding the initial singularity in classical general relativity without appealing to quantum effects. One could try to avoid the singularity by proposing that matter behaves in such a way in the very early universe that it does not obey the conditions laid down by

Hawking and Penrose. The most important of these conditions is a restriction on the behaviour of matter at high energies, called the 'strong energy condition'. There are various ways in which this condition might indeed be violated. In particular, it is violated during the accelerated expansion predicted in theories of cosmic inflation. Models in which this condition is violated right at the very beginning can have a 'bounce' rather than a singularity. Running the clock back, the universe reaches a minimum size and then expands again.

Whether the singularity is avoidable or not remains an open question, and the issue of whether we can describe the very earliest phases of the Big Bang, before the Planck time, will remain open at least until a complete theory of quantum gravity is constructed.

Time and Space, the Same but Different

The existence of a singularity at the beginning of the universe calls into question the very nature of space, and particularly of time, at the instant of creation. It would be nice to include at this point a clear definition of what time actually *is*.

Everyone is familiar with what time *does*, and how events tend to be ordered in sequences. We are used to describing events that invariably follow other events in terms of a chain of cause-and-effect. But we can't get much further than these simple ideas. In the end, the best statement of what time is, is that time is whatever it is that is measured by clocks.

Einstein's theories of relativity effectively destroyed the Newtonian concepts of absolute space and absolute time. Instead of having three spatial dimensions and one time dimension which are absolute and unchanging, regardless of the motions of particles or experimenters, relativistic physics merges these together in a single four-dimensional entity called space-time. For many purposes, time and space can be treated as mathematically equivalent in these theories. Different observers generally measure different time intervals between the same two events, but the four-dimensional space-time interval is always the same.

However, the successes of Einstein's theoretical breakthroughs tend to mask the fact that we all know from everyday experience that time and space are essentially different. We can travel north or south, east and west, but we can only go forwards in time to the future, not backwards in time to the past. And we are quite happy with the idea that both London and New York exist at a given time at different spatial locations. But nobody would say that the year 5001 exists in the same way that we think the present exists. We are also happy to say that what we do now causes things to happen in the future, but we don't consider two different places at the same time as causing each other. Space and time really are quite different.

On a cosmological level, the Big Bang certainly appears to have a preferred direction. But the equations describing it are again time-symmetric. Our universe happens to be expanding rather than contracting, but it could have been collapsing and be described by the same laws. Or could it be that the directionality of time that we observe is somehow singled out by the large-scale expansion of the universe? It has been speculated, by Hawking and others, that if we lived in a closed universe that eventually stopped expanding and began to contract, then time would effectively run backwards during the contraction phase. In fact, if this happened, we would not be able to tell the difference between a contracting universe with time running backwards and an expanding universe with time running forwards. Hawking was convinced for a time that this had to be the case, but later changed his mind.

Another, more abstract, problem stems from the fact that Einstein's theory is fully four-dimensional – the entire world-line of a particle, charting the whole history of its motions in space-time, can be calculated from the theory. A particle exists at different times in the same way that two particles might exist at the same time in different places. This is strongly at odds with our ideas of free will. Does our future really exist already? Are things really predetermined in this way?

These questions are not restricted to relativity theory and cosmology. Many physical theories are symmetric between past and future in the same way as they are symmetric between differ-ent spatial locations. The question of how the perceived asymmetry of time can be reconciled with these theories is a deep philosophical puzzle. There are at least two other branches of physical theory which raise the question of the 'arrow of time', as it is sometimes called.

One emerges directly from a seemingly omnipotent physical principle, called the Second Law of Thermodynamics. This states that the 'entropy' of a closed system never decreases. The entropy is a measure of the disorder of a system, so this law means that the degree of disorder of a system always tends to increase. I have verified this experimentally many times through periodic observation of my office. The Second Law is a 'macroscopic' statement – it deals with big things like steam engines – but it arises from a microscopic description of atoms and energy states provided by statistical mechanics. The laws governing these microstates are all entirely reversible with respect to time. So how can an arrow of time emerge?

Laws similar to the classical laws of thermodynamics have also been constructed to describe the properties of black holes and of gravitational fields in general. Although the definition of the entropy associated with gravitational fields is difficult to be precise about, these laws seem to indicate that the arrow of time persists even in a collapsing universe. It was for this reason that Hawking abandoned his time-reversal idea.

Another arrow-of-time problem emerges from quantum mechanics, which is again time-symmetric, but in which weird phenomena occur, such as the collapse of the wave-function when an experiment is performed. Wave-functions appear to do this only in one direction of time and not the other – but, as I have hinted above, this may well just be a conceptual difficulty arising from the interpretation of quantum mechanics itself.

The No-Boundary Hypothesis

Space and time are very different concepts to us, living as we do in a low-energy world far removed from the Big Bang. But does this mean that space and time were always different? Or, in a quantum theory of gravity, could they really be the same? In classical relativity theory, space-time is a four-dimensional construction wherein the three dimensions of space and one dimension of time are welded together. But space and time are still not entirely equivalent. One idea associated with quantum cosmology, developed by Hawking together with Jim Hartle, is that the distinctive signature of time may be erased when the gravitational field is very strong. The idea is based on an ingenious use of the properties of imaginary numbers. (Imaginary numbers are all multiples of the number i, which is defined as the square root of minus one.) This tinkering with the nature of time is part of the 'no boundary' hypothesis of quantum cosmology due to Hartle and Hawking. Since, in this theory, time loses the characteristics that separate it from space, the concept of a beginning in time becomes meaningless. Space-times with this signature therefore have no boundary. There is no Big Bang, no singularity, because there is no time, just another direction of space.

This view of the Big Bang is one in which there is no creation, because the word 'creation' implies some kind of 'before and after'. If there is no time, then the universe has no beginning.

Asking what happened before the Big Bang is like asking what is further North than the North Pole. The question is meaningless.

I should stress that the no boundary conjecture is not accepted by all quantum cosmologists. Other ways of understanding the beginning (or lack of it) have been proposed. For example, the Russian physicist Alexander Vilenkin has proposed an alternative treatment of quantum cosmology in which there is a definite creation, and in which the universe emerges by a process of quantum tunnelling out of nothing.

Theories of Everything

I have tried to describe just a few of the contributions that Hawking has made to the process, not yet completed, of welding together quantum physics with gravity theory. This, as I have tried to explain, is one step in the direction of what many physicists feel is the ultimate goal of science – to write the mathematical laws describing all known forces of nature in the form of one equation, one perhaps that you might wear on your T-shirt.

The laws of physics, sometimes also called the laws of nature, are the basic tools of physical science. They comprise mathematical equations that describe the behaviour of matter (in the form of elementary particles) and energy according to the various fundamental interactions described above. Sometimes, experimental results obtained in the laboratory or observations of natural physical processes are used to infer mathematical rules which describe these data. At other times, a theory is created first as the result of a hypothesis or physical principle which receives experimental confirmation only at a later stage. As our understanding evolves, seemingly disparate physical laws become unified in a single overarching theory. The examples given above show how influential this theme has been over the past 100 years or so.

But there are deep philosophical questions lying below the surface of all this activity. For example, what if the laws of physics were different in the early universe? Could one still carry out this work? The answer to this is that modern physical theories actually predict that the laws of physics do change. As one goes to earlier and earlier stages in the Big Bang, for example, the nature of the electromagnetic and weak interactions changes, so that they become indistinguishable at sufficiently high energies. But this change in the law is itself described by another law – the so-called electroweak theory. Perhaps this law itself is modified at scales where grand unified theories take precedence, and so on, right back to the very beginning of the universe.

Whatever the fundamental rules, however, physicists have to assume that they apply for all times since the Big Bang. It is merely the low-energy outcomes of these fundamental rules that change with time. Making this assumption, physicists are able to build a coherent picture of the thermal history of the universe which does not seem to be in major conflict with the observations. This makes the assumption reasonable, but does not prove it to be correct.

Another set of important questions revolves around the role of mathematics in physical theory. Is nature really mathematical? Or are the rules we devise merely a kind of shorthand to enable us to describe the universe on as few pieces of paper as possible? Do we discover laws of physics or do we invent them? Is physics simply a map, or is it the territory itself?

There is also another deep issue connected with the laws of physics, pertaining to the very beginning of space and time. In some versions of quantum cosmology, for example, one has to posit the existence of physical laws that exist, as it were, in advance of the physical universe they are supposed to describe. This draws many early universe physicists towards a neo-Platonic philosophy, in that what really exist are the mathematical equations of the (as yet unknown) Theory of Everything, rather than the physical world of matter and energy. On the other hand, not all cosmologists get carried away in this manner. To those of a more pragmatic disposition, the laws of physics are simply a useful description of our universe whose significance lies simply in their usefulness.

A Theory of Everything would consist of a further stage of unification of the laws of physics to include gravity. The main barrier to this final theory is the lack of any self-consistent theory of quantum gravity. Not until such a theory were constructed could it be unified with the other fundamental interactions. There have been many attempts to produce theories of everything, involving such exotic ideas as supersymmetry and string theory (or even a combination of the two, known as superstring theory). It remains to be seen whether such a grander-than-grand unification is possible.

However, the search for a Theory of Everything also raises interesting philosophical questions. Some physicists, Hawking among them, would regard the construction of a Theory of Everything as being, in some sense, reading the mind of God. Or at least unravelling the inner secrets of physical reality. Others simply argue that a physical theory is just a *description* of reality, rather like a map. A theory might be good for making predictions and understanding the outcomes of observation or experiment, but it is no more than that. At the moment, we use a different map for gravity from the one we use for electromagnetism or for weak nuclear interactions. This may be cumbersome, but it is not disastrous. A Theory of

Everything would simply be a single map, rather than a set of different maps that one uses in different circumstances. This latter philosophy is pragmatic. We use theories for the same reason that we use maps – because they are useful. The famous London Underground map is certainly useful, but it is not a particularly accurate representation of physical reality. Nor does it need to be.

And, in any case, one has to worry about the nature of explanation afforded by a Theory of Everything. How will it explain, for example, why the Theory of Everything is what it is and not some other theory? To my mind, this is the biggest problem of all. Can any theory based on quantum mechanics be complete in any sense, when quantum theory is in its nature unpredictable? Moreover, developments in mathematical logic have cast doubt on the ability of any theory based on mathematics to be completely self-contained. The logician Kurt Gödel has proved a theorem, known as the incompleteness theorem, which shows that any mathematical theory will always contain things that can't be proved within the theory.

Hawking in Perspective

So where does this leave us in assessing Hawking's place in physics and in wider society?

The first point to be made, and I make it meaning no disrespect whatsoever to Hawking and what he has achieved, is that it is absurd to compare him with Einstein and Newton. These characters ignited true revolutions in science and, in their different ways, the philosophical changes they brought about had great cultural impact. Stephen Hawking has not, by any stretch of the imagination, revolutionised his subject. His work has often been brilliant. His results have yielded important new insights into the way the universe works. He has developed new mathematical techniques and applied them to problems that nobody had tackled before. He is rightly regarded as one of the most able theoreticians of his day. But beyond that, the public image is out of all proportion to his place in the history of physical science.

A large part of this book has been devoted to the background to Hawking's work. In the course of this survey, the names of many great physicists were mentioned. In December 1999 the journal *Physics World* published the results of a poll of some of the world's leading physicists in which they were asked to name the five physicists who had made the most important contributions to the subject. In all, 61 names were mentioned on the lists received. The top of the poll was Einstein with 119 votes, followed by Newton with 96. Maxwell (67), Bohr (47), Heisenberg (30), Galileo (27), Feynman (23), Dirac (22) and Schrödinger (22) all appeared in the top ten. Only one of the 130 respondents put Stephen Hawking anywhere on his list. Yet, with the exception of Einstein, Newton and Galileo, none of those rated above Hawking is a household name.

Of course, there is an obvious additional factor in the case of Hawking. Nobody with a spark of humanity could fail to respond to his appearance without sympathy and admiration for his courage and resilience. When he was diagnosed with his illness in 1962, doctors gave him two years to live. Now, in 1999, he is 57 years old and still an active and productive research scientist. This is testament to a remarkable individual, but

I don't think the understandable human reaction to his plight is sufficient to explain his emergence as a media star of such immensity.

We can see clues to the origin of the Hawking phenomenon in the career of Albert Einstein. His intellectual achievements were clearly beyond the grasp of ordinary people, but this did not prevent him from becoming a world-wide media celebrity. In my book *Einstein and the Total Eclipse*, I explained how in Einstein's case, the media consistently placed him on the far side of a huge intellectual gulf that separated him from the common man, and people responded by treating him with the reverence usually reserved for the priesthood. People did not mind not understanding exactly what he did, but enjoyed believing that Einstein was an intellect greater than themselves.

I believe that much the same process has occurred with Stephen Hawking. Hawking too works in an area far remote from everyday circumstance, and deals with concepts counter to many common-sense notions. The huge sales of *A Brief History of Time* do not necessarily imply that Hawking's ideas are widely understood. I would even doubt whether the majority of those who have bought the book have ever read it. But Hawking's whole persona reinforces the 'other-worldliness' of his science. Even the strange artificial voice with which he speaks casts him in the role of a kind of oracle, speaking the secrets of the universe. The computer he uses to compose his speech makes it difficult for him to talk quickly. He has evolved a curious gnomic style to deal with this, which further adds to the sense of mystery. It is also difficult for him to take part in the to-and-fro of ordinary conversation. When Hawking speaks, you listen but don't interrupt.

Roughly 70 years separate the elevation of Einstein and Hawking to the cosmic priesthood. Although there are similarities, there are also differences. One is the role of the media itself. Einstein became famous before the days of television and radio. Nowadays, access to the media is much more rapid than it was in Einstein's day when newspapers were the main vehicles of mass communication. Instant access to the media tends to generate more noise and a stronger feedback, distorting and amplifying the popular significance of a person or event until its original status is lost.

Another change has been the development of technology to an extent unimaginable in Einstein's day. Not only mass communication, but also nuclear power, advanced electronics and computing, biotechnology and medicine have all developed dramatically over the last century. Technology has altered our lives in many ways, partly for the good, partly also in ways that have had a negative impact on society, producing alienation and resentment in some quarters. These days, there is a tangible backlash against those parts of science that impinge upon our daily lives through the technology they produce. Is it conceivable that, for example, a nuclear physicist or a biologist working on genetically modified food would be awarded the kind of recognition that Hawking has received? I think not. The very remoteness of Hawking's ideas from our everyday world removes any sense of threat from his science.

The Mind of God

To look at the development of physics since Newton is to observe a struggle to define the limits of science. Part of this process has been the intrusion of scientific methods and ideas into domains that have traditionally been the province of metaphysics or religion. In this conflict, Hawking's phrase 'to know the Mind of God' is just one example of a border infringement. But by playing the God card, Hawking has cleverly fanned the flames of his own publicity, appealing directly to the popular allure of the scientist-as-priest.

I am not by nature a religious man, but I know enough about Christianity to understand that 'knowing the mind of God' is at best meaningless and at worst blasphemous when seen in the context of that particular religion. But Hawking himself has been quoted frequently as saying that he does not believe in anything resembling the Christian God. Indeed, his notion of a world with no boundary (and hence no beginning and no end), described in all its aspects by a single mathematical 'Theory of Everything', has no place for a Creator at all. Hawking nevertheless believes that when (if) the Theory of Everything is discovered, it will explain 'whether the universe has a meaning, and what our role is in it', as well as enable us 'to know why the universe exists at all'. He thinks it possible to replace religion and metaphysics with a mathematical theory that encodes all the laws of nature. But the philosophical questions to be asked about the universe will inevitably involve some that cannot be answered in the framework of mathematics. Perhaps it will only be when a Theory of Everything is derived that physicists will realise that it falls short of this goal. Then, perhaps, cosmologists will begin to explore the metaphysical foundations of their subject more satisfactorily than they have done so far.

Further Reading

Essential reading for anyone interested in Stephen Hawking's science is his blockbusting popular work: Hawking, S.W., *A Brief History of Time*, Bantam Books, New York (1988).

At a higher technical level, there is also a collection of essays on various aspects of cosmology and astrophysics: Hawking, S.W., *Black Holes and Baby Universes and Other Essays*, Bantam Books, New York (1993).

A very nice illustrated exposition of the life and work of Stephen Hawking can be found in McEvoy, J.P. and Zarate, O., *Introducing Stephen Hawking*, Icon Books, Cambridge (1999).

The three main contributions of Stephen Hawking to relativistic cosmology and quantum gravity are presented in various papers in the technical scientific literature. A useful original source for his work on black hole evaporation is: Hawking, S.W., *Black Hole Explosions?*, Nature, **248**, 30 (1974).

The application of Roger Penrose's singularity theorems to the Big Bang is discussed in: Hawking, S.W. and Penrose, R., *The Singularities of Gravitational Collapse and Cosmology*, Proceedings of the Royal Society of London, **A314**, 529 (1970).

Quantum cosmology, including the no-boundary conjecture, is discussed in Hartle, J.B. and Hawking, S.W.,

The Wave Function of the Universe, Physical Review D., **28**, 2960 (1983).

For a simple outline of Einstein's general theory of relativity and the emergence of Einstein as a media personality, try: Coles, P., *Einstein and the Total Eclipse*, Icon Books, Cambridge (1999), to be re-issued in autumn 2000 as *Einstein and the Birth of Big Science*.

I think the best introduction to Einstein's theory and its various consequences for black holes and the rest is the following superb book by Kip Thorne: Thorne, K.S., *Black Holes and Time Warps: Einstein's Outrageous Legacy*, W.W. Norton & Co., New York (1994).

For an authoritative and beautifully written account of the search for theories of everything, see: Barrow, J.D., *Theories of Everything*, Oxford University Press, Oxford (1991).

For a vigorous polemic against the philosophical and religious claims of modern cosmologists, peppered with interesting historical and scientific insights, try: Jaki, S.L., *God and the Cosmologists*, Scottish Academic Press, Edinburgh (1989).

Paul Davies is one physicist who has pushed the view that the nature of modern physics indicates some form of design that allows life to exist within our universe. The most relevant of his many books is: Davies, P.C.W., *The Mind of God*, Penguin, London (1993).

Key Ideas

Newton's Mechanics

The theory of motion presented by Sir Isaac Newton in his great *Principia* (1686). It consists of a set of mathematical laws describing the rigidly deterministic motion of objects under the action of forces against the backdrop of an absolute space and absolute time. Newtonian mechanics governed the way in which scientists described the physical world for more than two centuries, until it was overthrown by experimental and theoretical developments in the early part of the 20th century.

Quantum Theory

Quantum theory describes the behaviour of matter on very small scales. The quantum world essentially comprises two distinct notions. One of these is that matter and energy are not smoothly distributed but are to be found in discrete packets called *quanta*. The other is that the behaviour of these quanta is not predictable as in Newton's theory, but that only probabilities can be calculated.

Relativity

Albert Einstein developed the theory of relativity in a series of monumental papers in the early part of the 20th century, beginning with the publication of the special theory of relativity in 1905 and culminating in the general theory of 1915. Relativity theory is a theory of space and time. It deprived physics of the absolute meaning of these concepts that was embedded in Newtonian mechanics. Dealing not with space and time separately, but with a hybrid concept called space-time (which can be curved and warped), relativity replaced Newton's law of gravity with a theory of how space can be distorted by the presence of mass.

Unified Theories

As physics has grown through the 20th century, it has brought more and more disparate phenomena within the scope of unified theories. The first major step in this programme was the unification of the theories of electricity and magnetism by James Clerk Maxwell, to produce a theory of electromagnetism. Theories now exist in which electromagnetism and the nuclear forces can be described in terms of a single set of mathematical formulae. Physicists would like to include the one force missing from this treatment so far – gravity – but this force has so far eluded attempts to include it. If and when gravity is unified, a 'Theory of Everything' would be the result.

Quantum Gravity

The 'missing link' in the chain of reasoning leading to a Theory of Everything is a mathematical description that combines the general theory of relativity with the ideas of quantum mechanics. Although much effort has been expended in the search for such a theory, formidable mathematical difficulties have defeated many attempts. Only in a few special cases have gravity and quantum theory been combined in an intelligible way.

Black Holes

Black holes are regions of space-time where the effect of gravity is so strong that light cannot escape. Black holes are thought to exist in nature, but though the evidence for them is compelling, it remains circumstantial. For theorists, black holes provide natural test cases in which to try to explore the consequences of fitting Einstein's general theory of relativity together with the principles of quantum mechanics. Hawking himself showed that quantum effects can allow black holes to radiate, so that they are not entirely black.

Singularities

A singularity is a point or region of space-time where the mathematical equations of a theory break down because some quantity becomes infinite. The centre of a black hole is an example of such a singularity in the general theory of relativity, as is the origin of the universe in the Big Bang model. Penrose and Hawking have proved a number of theorems about the nature and occurrence of these singularities. Their existence in Einstein's theory suggests that general relativity may be incomplete. A quantum theory of gravity is required to describe the properties of matter at the enormous densities that pertain at the Big Bang or in a black hole.

The Big Bang

The Big Bang is a term, originally coined by Sir Fred Hoyle, that describes the standard picture of the cosmos and how it evolves. Currently expanding and cooling, the universe was hotter and denser in the past. Clues to its high-energy phase can be found in its expansion, in the relic radiation that pervades all space, and in the trace quantities of light atoms cooked in the primordial nuclear furnace. The early stages of the Big Bang are used by particle cosmologists to study the character of the fundamental forces of nature. The Big Bang model breaks down at the very beginning of space and time because of the existence of a singularity. It is therefore seriously incomplete, and will remain so unless and until a quantum theory of gravity has been worked out.

Elementary Particles

The fundamental building-blocks of matter are called elementary particles. Modern particle theory classifies these particles into various types, according to properties such as mass, spin and electrical charge.

All particles found in nature can be described as combinations of a relatively small number of basic units. Examples of these basic units are the quarks which combine in various ways to form the heavy particles (protons and neutrons) that reside in atomic nuclei; and the leptons, an example of which are the electrons that orbit around the nucleus of an atom. There are three families of quarks and three generations of leptons. As well as stable atomic matter, quarks and leptons combine in various ways to make hundreds of unstable particles seen only in accelerator experiments.

There is also the fact that every particle (combination of quarks and/or leptons) also has a corresponding anti-particle, having the same mass but the opposite electrical charge. When a particle and its anti-particle meet, they annihilate in a burst of radiation.

The quarks and leptons combine with each other by virtue of interactions mediated by force-carrying particles called bosons. The force between charged particles is the electromagnetic force, and the boson that carries this force is the photon. Leptons interact among themselves and with other matter via the weak nuclear force, which is mediated by the W and Z bosons. Quarks interact with each other via the strong interaction, carried by bosons called gluons.

Other titles available in the Postmodern Encounters series from Icon/Totem

Derrida and the End of History
Stuart Sim
ISBN 1 84046 094 6
UK £2.99   USA $7.95

What does it mean to proclaim 'the end of history', as several thinkers have done in recent years? Francis Fukuyama, the American political theorist, created a considerable stir in *The End of History and the Last Man* (1992) by claiming that the fall of communism and the triumph of free market liberalism brought an 'end of history' as we know it. Prominent among his critics has been the French philosopher Jacques Derrida, whose *Specters of Marx* (1993) deconstructed the concept of 'the end of history' as an ideological confidence trick, in an effort to salvage the unfinished and ongoing project of democracy.

Derrida and the End of History places Derrida's claim within the context of a wider tradition of 'endist' thought. Derrida's critique of endism is highlighted as one of his most valuable contributions to the postmodern cultural debate – as well as being the most accessible entry to *deconstruction*, the controversial philosophical movement founded by him.

Stuart Sim is Professor of English Studies at the University of Sunderland. The author of several works on critical and cultural theory, he edited *The Icon Critical Dictionary of Postmodern Thought* (1998).

Foucault and Queer Theory
Tamsin Spargo
ISBN 1 84046 092 X
UK £2.99   USA $7.95

Michel Foucault is the most gossiped-about celebrity of French poststructuralist theory. The homophobic insult 'queer' is now proudly reclaimed by some who once called themselves lesbian or gay. What is the connection between the two?

This is a postmodern encounter between Foucault's theories of sexuality, power and discourse and the current key exponents of queer thinking who have adopted, revised and criticised Foucault. Our understanding of gender, identity, sexuality and cultural politics will be radically altered in this meeting of transgressive figures.

Foucault and Queer Theory excels as a brief introduction to Foucault's compelling ideas and the development of queer culture with its own outspoken views on heteronormativity, sado-masochism, performativity, transgender, the end of gender, liberation-versus-difference, late capitalism and the impact of AIDS on theories and practices.

Tamsin Spargo worked as an actor before taking up her current position as Senior Lecturer in Literary and Historical Studies at Liverpool John Moores University. She writes on religious writing, critical and cultural theory and desire.

Nietzsche and Postmodernism
Dave Robinson

ISBN 1 84046 093 8
UK £2.99   USA $7.95

Friedrich Nietzsche (1844–1900) has exerted a huge influence on 20th century philosophy and literature – an influence that looks set to continue into the 21st century. Nietzsche questioned what it means for us to live in our modern world. He was an 'anti-philosopher' who expressed grave reservations about the reliability and extent of human knowledge. His radical scepticism disturbs our deepest-held beliefs and values. For these reasons, Nietzsche casts a 'long shadow' on the complex cultural and philosophical phenomenon we now call 'postmodernism'.

Nietzsche and Postmodernism explains the key ideas of this 'Anti-Christ' philosopher. It then provides a clear account of the central themes of postmodernist thought exemplified by such thinkers as Derrida, Foucault, Lyotard and Rorty, and concludes by asking if Nietzsche can justifiably be called the first great postmodernist.

Dave Robinson has taught philosophy for many years. He is the author of Icon/Totem's introductory guides to Philosophy, Ethics and Descartes. He thinks that Nietzsche is a postmodernist, but he's not sure.